Mapping the Mountains

by Elizabeth J. Natelson

Printed in China

ISBN 978-0-15-362476-6
ISBN 0-15-362476-0

5 6 7 8 9 10   0940   16 15 14 13 12 11 10 09

Harcourt
SCHOOL PUBLISHERS

Visit *The Learning Site!*
www.harcourtschool.com

Introduction

If you plan to go hiking, bicycling, or horseback riding on a new trail, you might wonder what to expect. Is the path straight and level, or curving and hilly? If it's hilly, how high and how steep are the hills? Does the path pass mostly through shady forest or open meadowland?

You can't answer these questions by examining a globe. State highway maps or local street maps aren't useful for this purpose either. To learn about land features and elevation changes, you need a special type of map. Maps that show the elevation and features of the land are called topographic maps.

Topographic Maps

There are several ways to show the shape and elevation of landforms on a flat sheet of paper. One way is the shaded relief map, which is like a picture from an airplane. A different map, the topographic map, takes more practice to read but gives more detailed information. A topographic map shows the exact location and elevation of land features.

Most of the topographic maps that are used in the United States are made by the United States Geological Survey (USGS). The USGS is the agency of the federal government that is responsible for making maps and monitoring natural hazards, such as volcanoes.

Aerial Photographs

Topographic maps are often made from aerial photographs. These are pictures taken from a plane as it flies over the land. The plane flies in an exact pattern at just the right altitude.

Because landforms may look different on the ground, field workers check the photographs. These people also hike in the area that is being mapped. They compare what they see to what the photographs show. The field workers also research useful details, such as the names of landforms.

Finally, the information from the photographs is translated into colors, lines, and symbols on a topographic map.

This satellite image shows the topography, or land characteristics, of the Grand Canyon in Arizona.

Color-Coded

Topographic (or "topo") maps help you visualize the land you're traveling on. One way the maps do this is with colors. Different land features are shown with different colors.

On topographic maps forests are shown as green and water is shown as blue. White is used for open space, such as fields and prairie. Roads are shown as red or black.

By using the colors on a topographic map, you can quickly identify important land features.

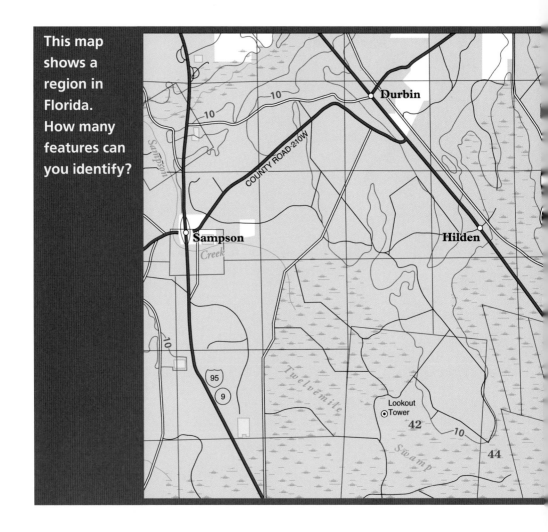

This map shows a region in Florida. How many features can you identify?

Other Marks

Topographic maps also show many other details, both natural and human-made. These features are identified with symbols. Waterfalls, glaciers, swamps, and caves are some natural features shown on topographic maps. Schools, churches, mines, power lines, and railroad tracks are examples of human-made features that are also included. The map's key explains each symbol on the map. A key is a listing of the symbols and the words that explain what each symbol means. Before using a topographic map, check the key for helpful information.

Contour Lines

The brown curving lines on a topographic map are called contour lines. *Contour lines* show changes in elevation, which is a measure of how high above sea level the land is.

To understand contour lines, look at the drawing below. It shows an island rising above the ocean. The level of the ocean represents 0 feet in elevation, or sea level. Imagine a line drawn all the way around the island at 100 feet above sea level, as shown in the diagram. This line is a contour line. Other lines can be drawn at 200 feet, 300 feet, and 400 feet. Each of these lines is also a contour line. Each contour line has the same elevation along its entire length.

The topography of this island can be shown with contour lines.

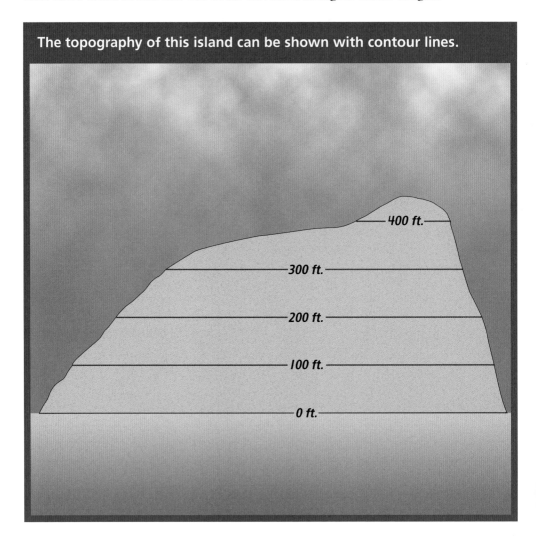

If you could look down at this sketch from above, the contour lines would appear nearly circular. Each contour line represents an elevation difference of 100 feet. This is how contour lines appear on a topographic map. Contour lines show how the land's elevation is changing.

The distance between the contour lines on a topographic map gives information about how steep the slope is. If the lines are close together, the land is steep. If the contour lines are far apart, the land is nearly flat.

Which side of the island has the steepest slope?

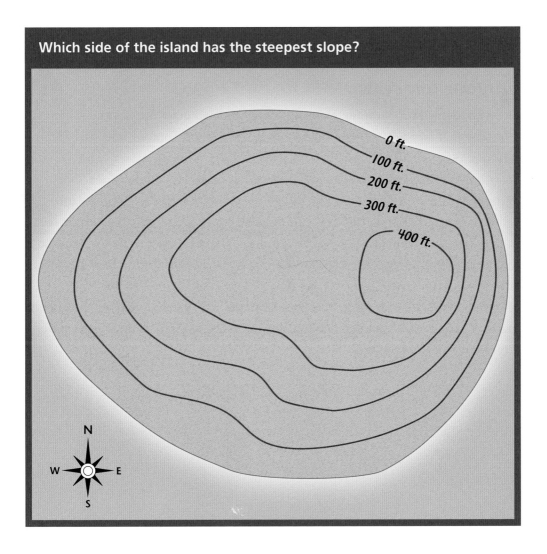

Index Contours

Every fifth contour line on a topographic map is a darker color. These lines are called index contours. An *index contour* has its elevation printed on it, also in darker color. The elevation is measured in feet above sea level.

Index contours are useful in two ways. One way is that they allow the map reader to determine how high the land is. Plain contour lines look the same whether they represent the top of a mountain or the floor of a deep valley. Index contours, with the elevation printed on them, clearly distinguish high places from low places.

Index contours also show whether the land slopes up or slopes down. Imagine that one index contour is 2000 feet and the next is 1800 feet. The land's elevation becomes lower toward the 1800-foot contour line.

Contour Intervals

Although all topographic maps have contours, all maps are not the same. One way that maps differ from one another is that their contour intervals are different.

Contour interval is the change in elevation between contour lines. On earlier pages, a contour interval of 100 feet was used to map an ocean island. Another contour interval that is commonly used on topographic maps is 40 feet. In this case, each contour line has an elevation 40 feet higher or lower than the line next to it. Knowing the contour interval helps you determine how steep the land is.

Each map's contour interval is printed in the map's white margin. In general, maps of flatter areas have smaller contour intervals. Maps of steeper areas have larger contour intervals.

Index contours are darker colored than other contour lines, and they have elevation labels. The contour interval is the change in elevation from one contour line to the next.

Contour interval = 40 feet

Scale

Topographic maps also have different scales. *Scale* is the relationship between the distance on a map and the distance on the ground. For example, many maps have a scale of 1:24,000. This can mean that one inch on the map represents 24,000 inches across the land. It can instead mean that one foot on the map represents 24,000 feet across the land. Any unit of measure can be used in a map scale.

Scale is important for calculating how far apart things are or how long a trail is. A map's scale is usually printed in the map's margin.

Hachures

Closed circles on a topographic map may represent either a hilltop or a hollow. To distinguish between these two features, some closed circles have hachures. A *hachure* (ha•SHUR) is a short line—one of many—pointing toward lower land. Closed circles with hachures inside represent a low point, not a hilltop.

Comparing Landforms to a Map

Many people use topographic maps when hiking. It's surprisingly easy to get lost on a trail and not know which direction to take. A map shows both the trail and the landforms around it. A hiker can compare the landforms with those shown on the map to decide which way to go.

Using a Compass

Comparing landforms to a topo map isn't always enough. Sometimes distinctive landforms aren't visible. However, a hiker may know enough to travel north or west. When a direction is all that's needed, a compass will provide it. A compass needle always points north, and that information is enough to allow a hiker to determine where the other directions are.

Using a Map and a Compass Together

A hiker might sometimes want to plan a route to a distant location. The location might be a park, town, or a particular landform. To do this, the hiker must use a map and a compass together.

First, the hiker can spot distinctive landforms that are nearby and find them on the map. This helps the hiker know where he or she is and which way is which. Next, the hiker uses the compass to find north. This direction can be compared to an arrow on the map that points north. The hiker can then orient the map in the proper direction. When the map's north arrow points in the same direction as the compass needle, the hiker can use the map to plan the most direct route to the location.

A compass needle points north. When the direction of north is known, the other directions can be read from the compass.

Magnetic North and True North

When using a compass and map, remember this: A compass needle points to magnetic north, but a map is oriented to true north.

Magnetic north is the direction toward Earth's magnetic north pole. The location of the magnetic north pole varies from year to year, but it is somewhere in northern Canada. *True north* is the direction toward Earth's geographic north pole.

The difference between magnetic north and true north is called *declination*. Declination changes, depending on where you are.

In St. Louis, Missouri, declination is nearly zero because magnetic north happens to be in line with true north. The declination of Boston is about 16° W. This means that the line from Boston to the magnetic north pole is 16° west of the line from Boston to the geographic north pole.

Some compasses can be adjusted for declination. Once adjusted, these compasses will indicate true north. For those that cannot, there is a simple solution. Suppose, for example, that you are near Seattle. The declination is about 18° E. You can hold the compass with its needle pointing at 18° east from north. This puts the 0° mark on the compass in the direction of true north.

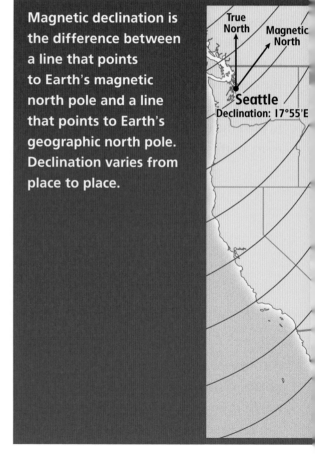

Magnetic declination is the difference between a line that points to Earth's magnetic north pole and a line that points to Earth's geographic north pole. Declination varies from place to place.

True North

Magnetic North

Seattle
Declination: 17°55'E

Every topographic map shows the magnetic declination at that location when the map was published.

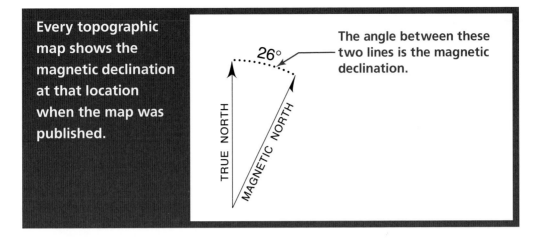

26°

The angle between these two lines is the magnetic declination.

TRUE NORTH

MAGNETIC NORTH

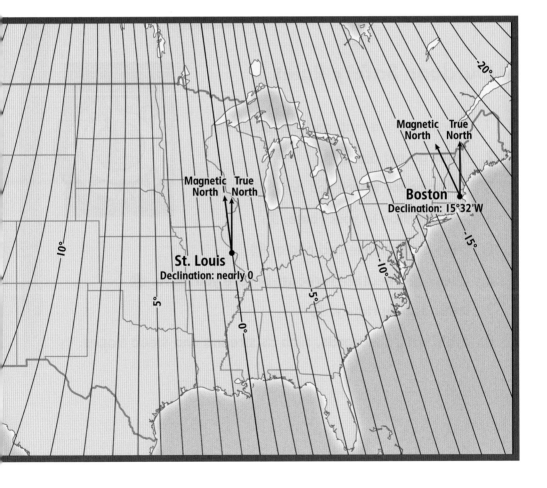

-20°

Magnetic North   True North

Boston
Declination: 15°32'W

Magnetic   True
North   North

St. Louis
Declination: nearly 0

-15°

-10°

10°

5°

-5°

0°

Maps and Safety

A map and a compass are fun to use. They make you less likely to get lost. However, using a topographic map takes practice. It won't help much to wait until you're lost and then pull out your map for the first time.

Having a map is only one of many important ways to stay safe in the outdoors. Here are three more: (1) never go out alone, (2) always let someone else know where you're going, and (3) actively pay attention to your route so that you know how to get back again.

Outdoor Safety Rules

1. Always carry a map and compass. ☺
2. Never go out alone. 🚫
3. Tell someone where you are going. ☺
4. Pay attention to your route. ☺
5. Bring a cell phone and radio. ☺
6. Check the weather forecast before you leave. ☺
7. Dress appropriately. ☺
8. Avoid natural hazards, such as streams and steep slopes. 🚫

Scientists Make Maps Everywhere

You might have seen maps of the ocean floor or of the surface of the moon. Technology allows mapmakers to chart mountains and valleys in places where they couldn't before. Most of these maps are made using a technique called remote sensing. Remote sensing means gathering data from a distance. The data can be acquired from a ship on the ocean or a space probe in orbit around a planet.

Maps of the Ocean Floor

Parts of the ocean floor have been mapped. Scientists send sound waves through the water using *sonar* (SOund Navigation And Ranging). The sound waves strike solid objects, including underwater mountains and valleys, and reflect back to the ship. The time it takes for the sound waves to reflect back tells scientists how deep the ocean floor is. Maps of the sea floor are then made from these data. Maps that show the depth beneath the ocean's surface are similar to topographic maps. But, the contours show depth instead of elevation.

Scientists have also made maps of planets and moons in the solar system. Instruments use various forms of radiation to collect data. Topographic maps of planets are made by using radar. Radio waves are sent to the planet's surface from an orbiting space probe. Scientists record the amount of time that passes before the waves are reflected back to the probe. This record provides information about the planet's topography. That data is then sent back to Earth, where scientists create computerized maps. This same technique can be used to map remote regions on Earth, such as Antarctica.

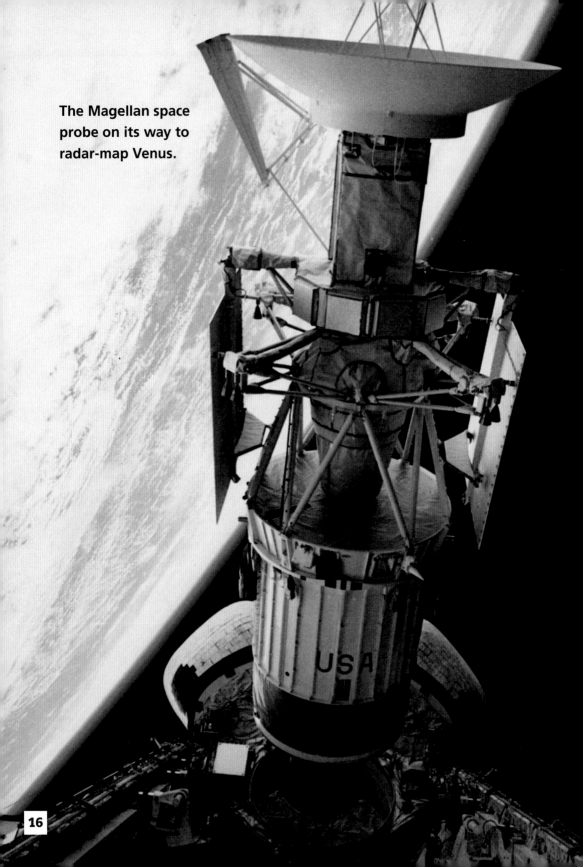

The Magellan space probe on its way to radar-map Venus.